ROBOT WORLD
ROBOTS IN THE FIELD

by Jenny Fretland VanVoorst

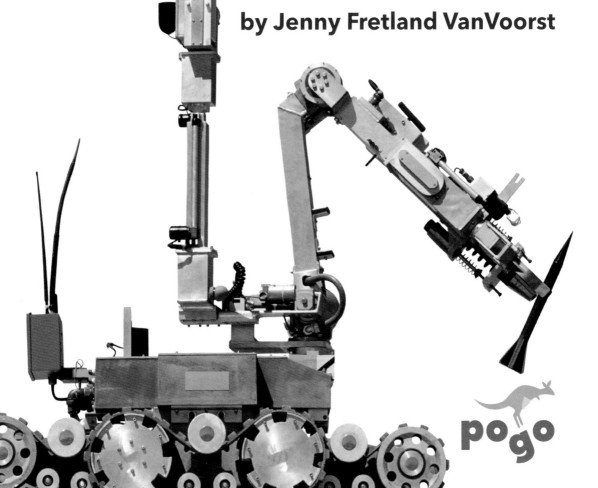

pogo

Ideas for Parents and Teachers

Pogo Books let children practice reading informational text while introducing them to nonfiction features such as headings, labels, sidebars, maps, and diagrams, as well as a table of contents, glossary, and index.

Carefully leveled text with a strong photo match offers early fluent readers the support they need to succeed.

Before Reading

- "Walk" through the book and point out the various nonfiction features. Ask the student what purpose each feature serves.
- Look at the glossary together. Read and discuss the words.

Read the Book

- Have the child read the book independently.
- Invite him or her to list questions that arise from reading.

After Reading

- Discuss the child's questions. Talk about how he or she might find answers to those questions.
- Prompt the child to think more. Ask: Can you think of other environments where robots could go more safely than people? What other dangerous jobs do you think robots could be used for?

Pogo Books are published by Jump!
5357 Penn Avenue South
Minneapolis, MN 55419
www.jumplibrary.com

Library of Congress Cataloging-in-Publication Data

Fretland VanVoorst, Jenny, 1972-
   Robots in the field / by Jenny Fretland VanVoorst.
      pages cm. – (Robot world)
   Includes bibliographical references and index.
   ISBN 978-1-62031-217-9 (hardcover: alk. paper) –
   ISBN 978-1-62496-304-9 (ebook)
   1. Robots–Juvenile literature.
   2. Robotics–Juvenile literature. I. Title.
   TJ211.2.F745 2015
   629.8'92–dc23
                                        2015020998

Series Designer: Anna Peterson
Book Designers: Anna Peterson and Michelle Sonnek
Photo Researchers: Anna Peterson and Michelle Sonnek

Photo Credits: Alamy, 10-11, 12-13; Carnegie Mellon University, 3; Corbis, 1, 6-7, 9, 20-21; DVIDS, 14-15; Getty, 17, 18-19; Science Source, 16; Shutterstock, cover, 8; SuperStock, 4; Thinkstock, 5, 23.

Printed in the United States of America at Corporate Graphics in North Mankato, Minnesota.

TABLE OF CONTENTS

CHAPTER 1

WHAT DO THEY DO?

Robots are a part of our world. They are in our factories and in our homes. They are even in space.

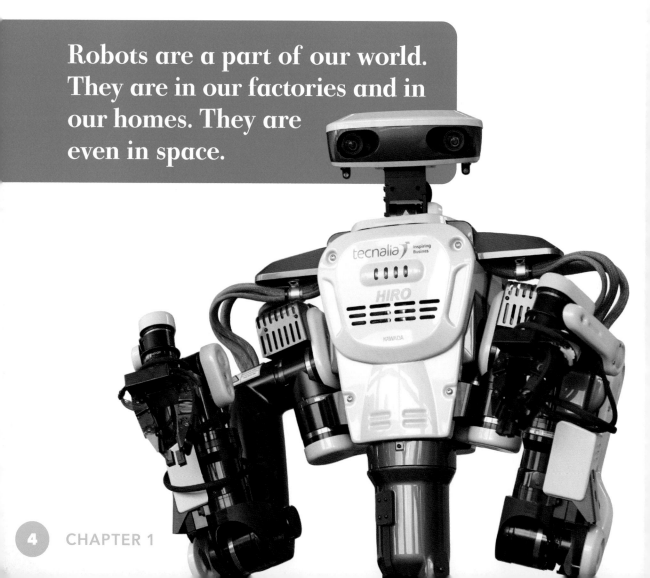

But they are also out in the field, doing jobs too dangerous for humans.

medical device robot

Robots **defuse** bombs and **land mines**. They carry soldiers' heavy loads. They search disaster sites to rescue survivors.

Robots work every day to make our lives safer. What are these machines? And how do they do it?

military pack
robot

CHAPTER 2

HOW DO THEY WORK?

In some ways robots are a lot like you. For example, how do you know to carry an umbrella when it is raining? You see or hear the rain. You decide an umbrella will keep you dry. You grab an umbrella.

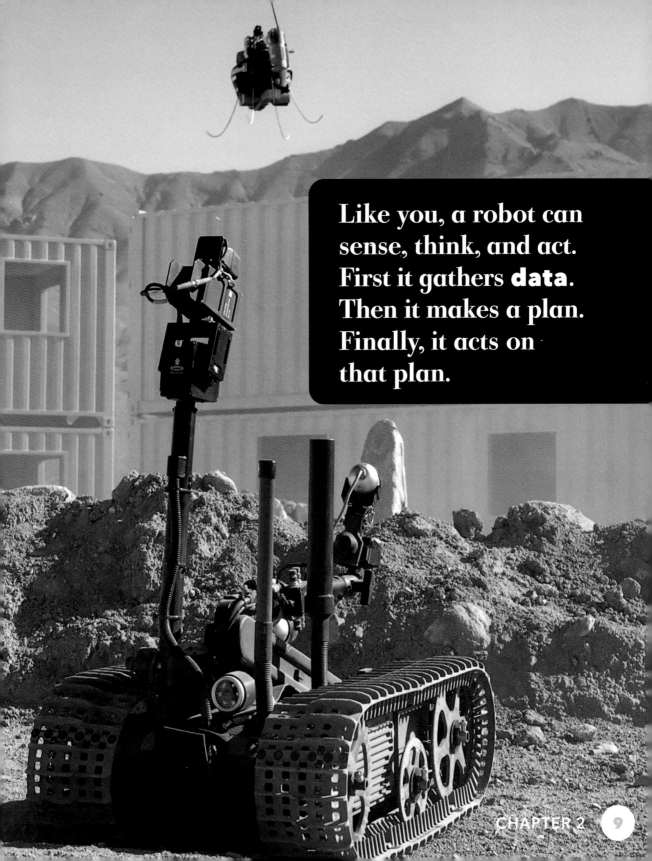

Like you, a robot can sense, think, and act. First it gathers **data**. Then it makes a plan. Finally, it acts on that plan.

sensors

Robots use **sensors** to gather data. Sensors help the robot understand its surroundings. They can be cameras and microphones. They can be heat or pressure sensors. They can be **GPS**.

TAKE A LOOK!

What makes a robot?

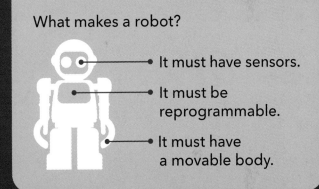

It must have sensors.

It must be reprogrammable.

It must have a movable body.

A robot is built for the task it is to perform. Traveling over rubble? It might have tracks instead of wheels. Searching tight spaces? It may be shaped like a snake and move like one too.

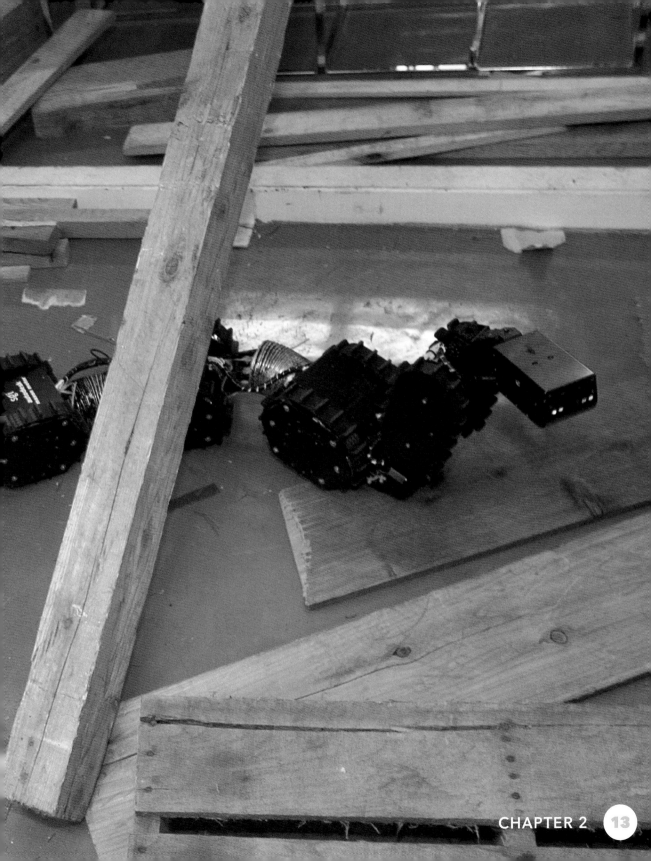

remote control

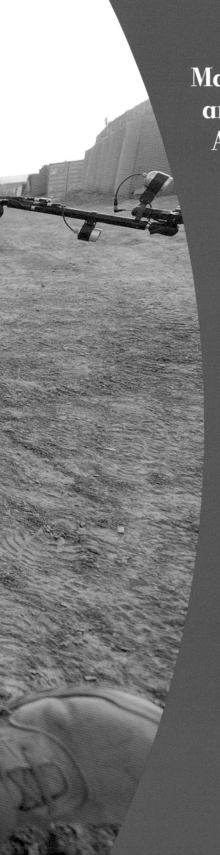

Many of these robots
are **remote controlled**.
A person takes in data
from the robot's sensors.
He decides what action
the robot should take.

CHAPTER 3

MEET THE ROBOTS

One use for robots in the field is in bomb disposal. This job could be deadly for a human. But if a robot is destroyed, it's OK. Another one can be built just like it.

A human operator controls the robot. She might use the robot to safely move a bomb or cut a wire to disable it.

Robots are also used for **search and rescue** missions. Disaster sites are dangerous. They are also dark and dirty. People can't get into the rubble. They can't see in the dark or in the smoke or dust. But a robot can.

Robots see in the dark using **lidar**. They use heat sensors to detect body heat. They use microphones to listen for voices.

Robots help us in so many ways. But robots in the field go even further. They are built to save our lives!

DID YOU KNOW?

In 2015 the U.S. government held a robotics contest. It focused on search and rescue. Robots had to open a door. They climbed a ladder. They used tools to break through a wall. They even had to drive a car!

ACTIVITIES & TOOLS

TRY THIS!

ROBOT PARTS

Believe it or not, a lot of the parts that make up a robot are things you can find right where you live.

Take a tour of your home and see how many of of the following robot components you can find. Ask an adult if you need help.

Look for a:

- Switch or button
- Motor
- Indicator light
- Programmable device (such as a DVR)
- Wheel
- Gear
- Heat sensor (such as a thermometer)
- Light sensor (such as a night light)

GLOSSARY

data: Facts about something that can be used in calculating, reasoning, or planning.

defuse: To remove the part of a bomb that makes it explode.

GPS: A navigation system that uses satellite signals to find the location of a radio receiver on or above the earth's surface; abbreviation of global positioning system.

land mines: Bombs buried in the ground that explode when they are stepped or driven on.

lidar: A device that uses laser beams to detect and locate objects.

remote control: To control the operation of something from a point some distance away.

search and rescue: The process of seeking out and providing aid to people who are in distress or danger.

sensors: Onboard tools that serve as a robot's eyes, ears, and other sense organs so that the robot can create a picture of the environment in which it operates.

INDEX

TO LEARN MORE

Learning more is as easy as 1, 2, 3.

1) Go to www.factsurfer.com

2) Enter "robotsinthefield" into the search box.

3) Click the "Surf" to see a list of websites.

With factsurfer, finding more information is just a click away.